50

WAYS TO
SAVE
WATER
& ENERGY

SIÂN BERRY

ABOUT THE AUTHOR

Siân Berry is the Green Party's candidate for London Mayor in 2008 and is a founder of the successful campaign group, the Alliance Against Urban 4x4s.

Siân was one of the Green Party's Principal Speakers until September 2007 and was previously national Campaigns Co-ordinator. She stood in the Hampstead and Highgate constituency in the 2005 General Election and has campaigned in her local area for more affordable housing and, nationally, to promote renewable energy and local shops.

Famous for the mock parking tickets created by Siân in 2003, the Alliance Against Urban 4x4s is now a national campaign and the group recently celebrated persuading the current Mayor of London to propose a higher congestion charge for big 4x4 vehicles and other gas-guzzlers.

Siân studied engineering at university and her professional background is in communications. These skills give her a straightforward and accessible approach to promoting green issues, focusing on what people can do today to make a difference, and on what governments need to do to make greener lives easier for everyone.

As spokesperson for the Alliance and a well-known Green Party figure, Siân has received wide coverage in national and international newspapers and has appeared on numerous TV and radio shows, from Radio 4's *Today* programme to *Richard and Judy*. Her calm, cheerful and persuasive advocacy has stimulated a lively public debate about 4x4s, and has helped to raise the environment further up the public agenda.

50

WAYS TO
SAVE
WATER
& ENERGY

SIÂN BERRY

Kyle Cathie Ltd

First published in Great Britain in 2007 by
Kyle Cathie Ltd
122 Arlington Road, London NW1 7HP
general.enquiries@kyle-cathie.com
www.kylecathie.com

10 9 8 7 6 5 4 3 2 1

978-1-85626-773-1

Editorial Director: Muna Reyal
Illustrator and Designer: Aaron Blecha
Production Director: Sha Huxtable
Junior Editor: Danielle Di Michiel

A Cataloguing In Publication record for this title is
available from the British Library.

Colour reproduction by Scanhouse
Printed and bound in Italy by Amadeus

Printed on 100% recycled paper

**THE MAN WHO MOVES
A MOUNTAIN BEGINS
BY CARRYING AWAY
SMALL STONES**

Confucius

50 WAYS TO...

In the home, in the garden, at the shops, at work and on the move, this series of books contains a wide range of simple ways to live a greener life, whatever your situation. Each book has 50 easy, affordable and creative tips to help you live more lightly on the planet.

There are many ways to be green that don't need a big investment of money, time, effort or space. Saving energy also saves money on your bills, and eco-friendly products don't have to be high-tech or expensive.

Any size garden – or even a window box – can be a haven for wildlife and provide useful low-maintenance crops that save on imported fruit and vegetables. And those of us living in towns and cities should know that urban living can provide some of the lowest-carbon lifestyles around.

The 50 Ways series has been written by Siân Berry: Green Party candidate for Mayor of London in 2008 and a founder of the successful campaign group, the Alliance Against Urban 4x4s. She shares her experiences to demonstrate how you can reduce your carbon footprint, stay ahead of fashion and enjoy life without sacrifice.

Siân says: 'Being green is not about giving everything up; it's about using things cleverly and creatively to cut out waste. In these books, I aim to show you that a greener life without fuss is available to everyone'.

INTRODUCTION

Saving water and energy sounds like a good thing to do. But why is it so important?

Unlike other kinds of waste, energy is invisible. We don't have to put the energy we waste in a bin for someone to collect, and waste water helpfully disappears down the drain, where we no longer have to think about it.

But, even if we can't see the problem, water and energy are the most precious resources in the world and wasting them also affects our pockets, not just the environment.

Where does our water come from? How much are we using, and for what? What sources of energy do we use now, and how long will they last?

These are important questions, so before we get into the nitty gritty of saving water and energy in our daily lives, let's have a look at some of the basics.

Water

The Earth is full of water, but more than 99 per cent is either salt water in the sea or frozen water in ice caps and glaciers. Only 0.3 per cent of the world's water is available for use by humans, plants and animals to support life on land.

Our supplies of fresh water largely come from the ocean, where it evaporates into clouds and eventually falls as rain. Where the rain falls depends on weather patterns, and by encouraging climate change, we are changing these patterns in ways that are hard to predict. We have seen only a small change in our weather so far, but still there have been devastating floods and severe droughts across the world, both of which affect water supplies.

Saving water also saves energy. The water from your taps uses energy twice: first to clean and purify it; and second, to treat it when goes down the drain. We also use a lot of energy to heat water in our homes, so not wasting hot water is even more important.

Finally, saving water can even save you money. Once you are practising water conservation and using less than the average home, ask for a water meter to be fitted, and your bills will actually go down!

Energy

Most of the energy on the Earth comes from the heat and light of the sun, but there is also energy in molten rocks inside the Earth, as well as energy in the tides created by the gravity of the sun and moon.

However, at the moment, almost all the energy used by humans comes from fossil fuels, such as oil, coal and gas. These are the ancient remains of plants that captured energy from the sun millions of years ago.

There are two problems with burning fossil fuels. The first is that they aren't going to last. It won't be long before we can no longer increase the rate we pump oil from under the ground to meet demand. Gas reserves are limited, and even the vast deposits of coal around the world can't be mined for ever.

The second – and most serious – problem is all the carbon dioxide released by burning fossil fuels. As it builds up in the atmosphere, carbon dioxide causes higher temperatures by enhancing the greenhouse effect, and this is already starting to have an impact on our weather.

Climate science tells us that we must reduce our worldwide carbon dioxide emissions dramatically over the next few decades in order to prevent runaway climate change. Countries that are the worst polluters must make the biggest cuts if the world is going to meet this target. So, we must start now by reducing our energy needs and moving to new ways of harnessing energy that don't release carbon dioxide.

This book concentrates on saving energy. The energy we currently produce could serve our needs many times over if we used it better, and reducing waste energy costs much less than producing more.

The energy we use can be divided roughly into three basic uses: one third by industry, one third by transport and one third in the home.

There is a lot we can do as individuals to reduce energy used for transport, and by shopping carefully and using our 'consumer elbow' to encourage businesses to be greener, we can help there, too. Pressure on politicians is also important, because they can make international agreements, set energy policy and change the rules for how businesses work.

In this book, I will look at the portion of energy we use in the home. There are lots of easy ways we can save energy every day, and by doing this, we can make a real contribution to reducing climate change, saving ourselves money at the same time. After all, the cheapest energy is energy that doesn't need to be paid for at all, because we don't need it anymore!

SAVING WATER

Fresh water is scarce and precious. Over a billion people don't have access to clean water and this number is increasing. By 2025, two thirds of the people on Earth could be living in conditions of 'water stress' and many of these will be in rich countries, not just in areas traditionally linked with droughts.

The average person uses hundreds of litres of fresh water a day in the home. But the overall water consumption we are responsible for across the world is more than twenty times higher. The food and goods we buy consume water on their way to us and this 'hidden' water really adds up.

Our use of water has an impact globally and can affect countries, habitats and people that are very short of water. The chapter on hidden water will show how you can reduce the impact of the goods and food you buy on global water resources.

The water we use in our homes has the most impact on water resources in our own countries. We can't easily pipe in water from other areas and, unlike oil used for energy, there is no substitute for clean water once it has been polluted or sent out to sea.

In the first section of this book, I will be looking at lots of simple and clever ways we can save water every day – by using it better in the home and by shopping more wisely.

Of course, there's one area where we shouldn't save water, and that's when we drink it – healthy living depends on drinking plenty of fresh water, and we should all be doing more of this, not less!

WASHING OURSELVES

You may think that the toilet is the biggest water-waster in the house, but it actually comes in second place.

The water we use for washing ourselves, including hand-washing under a tap, cleaning our teeth, showering and taking baths, uses up about a third of our home water supply.

These activities are a big source of waste, too – simply letting a tap run can send 20 litres down the drain in just one minute. There are some very easy ways to save water with better fittings and with better washing habits.

HAVE A SHOWER NOT A BATH

The average bath uses up one hundred litres of water, whereas a quick shower uses up just thirty litres, so this is a really easy way to save water on a daily basis.

You may be saving energy by having your water heated by solar panels, but having a bath will still waste water on a grand scale, so keep baths for special occasions (such as the day you get back from a camping trip) and have regular showers instead.

GET A NEW SHOWER HEAD 2

Power showers are a popular choice in many homes. These are showers that use an electric pump to increase the water flowing through the shower head. Spend more than a few minutes under one of these and you could go through the same amount of water as you do in the bath.

Whatever type of shower you have, you can save more than half the water it uses and still get the same shower experience by simply changing your shower head.

An aerating shower head mixes air into the flow to keep the pressure high but reduce the water used. The bubbles feel great and actually increase the amount of water in contact with your body by four times.

3 SORT OUT YOUR TAPS

It goes almost without saying that leaky, dripping taps should be fixed as soon as the leak develops.

The amount of water that can be wasted this way could come as a surprise. Even a leak that drips only once per second adds up to more than 15,000 litres in a year and a thin trickle could mean 100,000 litres of water going straight down the drain.

Washing your hands under a fast-running tap is another way of wasting a lot of water, because the amount that actually cleans your hands is a tiny fraction of what comes out of the tap.

The pressure of water in the mains pushes up to twenty litres of water per minute through a fully open tap, but you can wash your hands just as effectively in water that flows at six litres per minute. Try running the tap at a lower flow rate and you'll see for yourself.

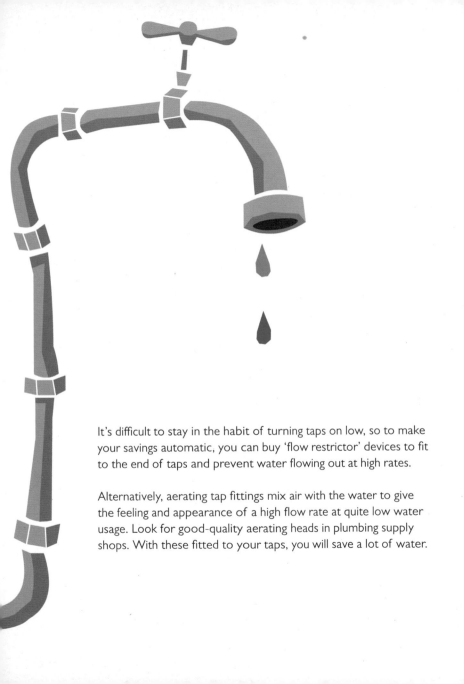

It's difficult to stay in the habit of turning taps on low, so to make your savings automatic, you can buy 'flow restrictor' devices to fit to the end of taps and prevent water flowing out at high rates.

Alternatively, aerating tap fittings mix air with the water to give the feeling and appearance of a high flow rate at quite low water usage. Look for good-quality aerating heads in plumbing supply shops. With these fitted to your taps, you will save a lot of water.

SAVING WATER IN THE LOO

It's no surprise to find that the toilet is responsible for 30 per cent of home water consumption. Each of us flushes the toilet more than a thousand times a year, so paying attention to the water consumed by your loo is an area where great savings are possible.

While not flushing isn't really an option (unless you want to follow the principle of 'if it's yellow, let it mellow'), there are many new products around to reduce the water used each time you flush.

And the toilet should never be used as a waste bin. Flushing away anything other than natural waste and toilet paper doesn't just waste water. Other items can cause blockages in your pipes or at the sewage plant, and can end up being sent out to sea where they are a hazard to wildlife, before ending up as litter on beaches.

Plasters, cotton buds, wipes, condoms, contact lenses and sanitary products should all be wrapped and put in the bin, not the toilet.

4 GET THE RIGHT CISTERN

Toilets vary hugely in the amount of water they use per flush. Pre-1950s models can use up to thirteen litres, whereas modern cisterns only hold about six litres.

Newer toilets also have a dual-flush option, with a smaller button that can be used when there is less to flush away. The half-flush option uses around half the water of a full flush, so use this button wherever possible.

WATER-SAVING DEVICES 5

Whatever you do, don't follow the old advice about putting a brick in the cistern. Over time, a brick will break down and release abrasive particles, which can damage the mechanism, and even cause catastrophic leaks.

If you have a large, old toilet, you can still save a litre or more of water every time you flush by adding a modern, plastic water-saving device to the cistern.

These come in various forms. The simplest water saver, the Hippo, is just a stiff plastic bag that sits upright in the cistern to retain some of the water when it empties.

You can even improvise and make your own water-saving device by filling a 1-litre plastic bottle right to the top with water, doing the lid up tightly and putting it in your cistern. Make sure there are no air bubbles or the bottle will float.

Dual-flush or variable-flush handles, which give the option of a shorter flush, can be fitted to older loos without having to replace the whole system. These are available from DIY shops, or ask your plumber to fit one next time your toilet needs repairing.

6 FIX A LEAKY LOO

Leaks between the cistern and the toilet bowl can also increase water waste. If you have a really bad leak, you will be able to see a constant trickle of water running down the back of the bowl.

Smaller leaks can be spotted using a bit of simple detective work. Add a few drops of food colouring to the cistern. If you can see the colour in the bowl after an hour or two, you've got a leak, so call in a plumber to fix it.

FLUSH WITH GREY WATER 7

Making water clean enough to drink takes lots of effort. Many different processes that use up energy and chemicals take place at the water treatment plant, yet only about four per cent of the fresh water in the average home is used for drinking and cooking. Most is used for purposes where drinking quality isn't required.

Grey water is the name for rainwater or water that has previously been used for tasks such as washing. It can be used for jobs that don't need clean water, such as flushing the toilet. Grey-water systems are now compulsory for new houses in some areas of Germany.

Grey-water systems in the home can range from a simple tank to collect water from sinks to use in the toilet, to a fully integrated rainwater and grey-water collection, treatment and piping system.

In the future, sophisticated ways of reusing grey water in the home will be built into new homes. If you are building a house, or simply refitting your bathroom, find out if you can take the opportunity to use grey water in some way. Replacing just some of the clean water you use to flush the toilet could mean big savings.

The easiest place to reuse grey water is outside the home. See the section on saving water in the garden for more details.

IN THE LAUNDRY

If you don't have a washing machine at home and use the launderette instead, a gold star to you! Communal laundries have been shown to use two-thirds less water per person served than individual washing machines in every home.

The reasons behind this saving include the fact that having a machine at home encourages us to use it more often and for smaller loads, as well as the fact that launderette machines tend to be better maintained and more efficient in their use of both water and energy.

I get my washing done at my local launderette, mainly because of lack of space and the trouble of having a machine installed in my new flat. As a result I only do a wash once or twice a week, while the average washing machine in the home does 270 washing cycles a year. I even think I will save money in the long term; in my previous houses, getting washing machines repaired was an expensive business, and I've suffered more than one inconvenient and destructive kitchen flood.

However, the launderette isn't for everyone. If you have a large family and need to wash clothes more often, it makes sense to have a machine at home. And there are still ways to save water by choosing the right machine and using it wisely.

8 ECO-FRIENDLY MACHINES

All washing machines have an energy label displayed in the shop. This gives the machine a score for energy efficiency and also gives a figure for water consumption.

Older washing machines used up to one hundred litres per washing cycle, but the average for new machines is around fifty litres, while the very best can use as little as thirty litres per wash.

Look out for the European Eco-label (a flower with the European stars around it). This certifies that a washing machine will save both water and energy, and it will be manufactured in a more eco-friendly way.

If you can, choose a machine that is also A-rated for its spin cycle, because this means you will get much drier clothes at the end of the wash, avoiding the need to tumble dry and speeding up line or hanger drying.

FILL IT UP 9

The water consumption written on the energy label is a good guide to the water you can save by buying an eco-friendly machine. But this only really counts if you are filling up the machine with as many things as possible in each wash, reducing the water used per piece of clothing as well.

Clothes don't need 'room to breathe' in modern washing machines. In fact, you'll only get the full, A-rated benefits if you put in a full load every time. Washing smaller amounts and pressing the half-load button saves about a quarter of the water and energy of a full cycle, so this isn't a real water-saving option. Far better to wait until you have a full load of washing to do.

See Tip 41 for more on why running a full load is a good idea.

10 USE ECO-DETERGENTS

All detergents contain molecules that attract water at one end and grease at the other. This is how they help to wash away dirt from your clothes.

Most washing powders contain detergents made from petrochemicals derived from oil, as well as a lot of other chemicals, including perfumes, bleaches and enzymes. There is no legal requirement for all these ingredients to be biodegradable, so large amounts of chemicals and additives end up being washed into our rivers and streams, where they can affect wildlife.

Instead of going for the big-brand washing powders, look for products that are made from renewable resources, such as plants, and which are fully biodegradable. These are just as good as the brands that do all that advertising, and using them will help save fresh water for other, more useful, purposes.

WASHING-UP

Whether to wash dirty dishes by hand or get a dishwasher is a difficult question. I have done both while living in different houses, and given the huge amounts of water I could hear sloshing around inside the dishwasher (and the clouds of hot steam that come out at the end), I believed for a long time that hand-washing was much better for the planet.

However, technology has moved on and the best dishwashers use less water now than the typical hand-washing routine – as little as sixteen litres per wash versus forty to wash and rinse a dozen place settings by hand. If you prefer not to get your hands wet, there are now plenty of eco-friendly dishwashers out there to choose from.

Of course, the resources used by the machine aren't the only thing to take into account. Lots of energy and water are consumed in the process of manufacturing and transporting dishwashers to the shops, so if you bear in mind some green principles when washing by hand, you can still do better overall.

ECO-WASHING BY HAND

Don't run the water too hot. This is bad for your hands as well as the planet, and you will also risk creating stress cracks in your glasses. I have a thing for vintage, coloured glasses from flea markets, and have lost several of these thanks to carelessly plunging them into very hot water.

Resist the urge to rinse everything under a running tap. Even a relatively thin trickle wastes around five litres a minute, significantly reducing the benefits of washing by hand. Use ecological washing-up liquid and there will be fewer toxic residues to rinse away.

If you do feel that rinsing is essential, wait until you have piled everything up in the drainer and very slowly pour a couple of glasses of clear water over everything. This will use far less water than keeping a tap running for half an hour while you rinse things one by one.

RUN FULL LOADS 12

As with washing machines, the efficiency of dishwashers is highest when they are run with a full load of dirty dishes inside. Always scout around the house for every plate and cup that needs doing before turning on your machine and you'll make the most of its eco-credentials.

And never wash anything twice. Rinsing all your plates under a tap before putting them in the dishwasher is a guaranteed way to waste water. Instead, simply scrape any leftovers into the bin (ideally your compost bin) before loading into the machine.

COOKING

Even if you are the keenest environmentalist in the world, there's no need to save water by reducing the amount you drink or cutting down on the water in your food.

But, in the kitchen, plenty of water does go to waste in the process of cooking and preparing our meals. A lot of this is easily avoided and these tips can even help to improve the taste of your food, and keep vitamins and other healthy things inside.

13 WASH VEG IN A BOWL

Chemical residues on our fruit and veg, as well as the need to remove soil and other dirt, can mean that a lot of ingredients need to be washed before cooking.

We already know how a running tap can waste a hundred litres of water in a few minutes, so scrubbing a bag of potatoes could lead to a huge amount of fresh water going down the drain.

The best way to clean vegetables, salads and fruit is in a bowl of water, rather than under the tap. For lettuce, cabbage and other green leafy vegetables, this means a lot more water gets into the ridges and crevices in the leaves, which actually gets them cleaner than a quick rinse under running water.

Invest in a mechanical salad spinner for easy drying and you'll have clean, healthy salads without pouring huge quantities of water away.

REUSE WATER 14

OK, it's true that reusing water in the kitchen only saves a small amount of water, but it has other benefits that make it a very worthwhile thing to do.

The water used to steam or boil vegetables contains a lot of flavour, as well as all the goodness from those water-soluble vitamins that escaped into the water during cooking – so don't throw it away. This basic vegetable stock can be used to make soups or top up casseroles, and will improve the flavour of other dishes you are cooking at the same time.

You can also boil more than one thing in the same pan: try putting eggs in with peas to save space on the stove, not to mention the water and energy needed to boil an extra pan.

15 STEAMING IS BEST

Boiling vegetables in a big pan isn't a great use of water, and it isn't particularly healthy either. Essential vitamins and minerals escape into the boiling water, and the texture of over-boiled veg is none too appetising. I can remember thinking soggy cabbage was normal because that's how it always was at the school canteen.

Now, steamed cabbage is one of my favourite things, thanks to my stack of bamboo steamers. Steaming vegetables with these is really easy. Hardly any water escapes, so they don't need lots of attention, and I can stack up different ingredients on top of each other to cook at the same time, saving space on the stove as well.

The texture and taste of steamed new potatoes is so much better than when they are boiled that I wouldn't go back to boiling veg now if you paid me.

Steamed vegetables are healthier, too. Broccoli that is microwaved or boiled in water loses most of its beneficial antioxidants, but almost 90 per cent of these antioxidants are preserved when it is steamed instead.

CUT DOWN ON HIDDEN WATER

Almost everything we buy or use consumes fresh water while it is being made. The impact of this hidden water often isn't felt in our immediate neighbourhoods, but it extends right around the world. By buying cotton products from Egypt, we are effectively importing water from the Nile. And in some African countries, water shortages are being created because of the amount of irrigation water applied to roses grown for export.

Food production is the heaviest consumer. Only a fraction of crops is watered artificially from rivers and underground stores, but 70 per cent of fresh water used worldwide is dedicated to this purpose.

It takes around 1,000 litres of water to produce a kilogram of wheat, and a single cup of coffee can have a water footprint of 140 litres, while meat production is a very water-intensive way of feeding ourselves.

Other products, such as cars and fridges, aren't an obvious source of water waste but, in many industrialised countries, the amount of river water and groundwater extracted for factories and homes is higher than the amount used for agriculture.

The trade in hidden water isn't measured carefully in the same way that carbon dioxide emissions are, but studies have shown that it is enormous: a mind-boggling one thousand billion cubic metres of water flows around the world in traded goods every year.

There is now a website where you can enter information, such as where you live and what you eat, to calculate your approximate 'water footprint'. Visit www.waterfootprint.org and you'll be amazed at how it all adds up.

It is difficult to know how much hidden water you can save by changing your habits, but there are some easy steps that will definitely make a difference.

16 BUY LOW-WATER GOODS

This isn't as simple as it might sound. Whether a particular item is a heavy user of water depends on how and where it is produced, so a league table of 'best products' would be very complicated.

However, there are some basic principles that can help to make water-saving choices simpler.

Plants grown in their natural season need less artificial watering so have a much lower water footprint.

Organic farming saves water, because a lot of water is needed to dilute and apply pesticides and fertilisers to non-organic crops.

Different varieties of the same foods also have different water needs. For example, Desirée potatoes are a better choice than Maris Piper potatoes, which need more water to produce a good crop (and a better choice for gardeners wanting to save water, too!).

For industrial products, clothes and other goods, hidden water is very hard to measure. Reducing the number of products you buy, repairing things to make them last longer, and finally, recycling or reusing them all helps to cut down on wasted water.

By being careful about the amount of material you send to landfill, you are also preserving resources and energy, so it's nice to know you can add 'water saving' to the long list of benefits of an eco-friendly, waste-free lifestyle.

BE MORE VEGETARIAN 17

One thing that definitely uses up more water than necessary is the farming of animals for meat.

Kilogram for kilogram, beef uses fifteen times more water to produce than wheat, thanks to the crops grown to feed the cattle as well as the water the animals actually drink. Because of its water-intensive production, the water footprint of one beefburger is a massive 2,400 litres.

If you don't want to give up meat entirely, becoming a 'semi-vegetarian' is surprisingly easy – and healthier, too. Studies have shown that a vegetarian diet is generally lower in fat and higher in vitamins and other nutrients than a meat-eating diet. Vegetarians spend less time in hospital compared with meat-eaters, and suffer less from heart disease, high blood pressure and bowel diseases.

I'm not completely vegetarian, but over the years, I have added plenty of vegetarian and vegan dishes to the range of meals I cook. I also try to avoid factory-farmed meat when I eat out, so my meat consumption is very low – some weeks I don't eat any meat at all.

I know this is better for me, as well as for the environment as it saves water and energy. I also find that I eat more interesting food. Middle Eastern and Asian restaurants have a huge range of delicious meat-free dishes, which I missed out on when I used to ignore the vegetarian part of the menu.

SAVE WATER IN THE GARDEN

Water used in the average garden is about seven per cent of the total consumed by households. But during a hot summer, when sprinklers and hoses are brought out, this rises to more than half.

Apart from the complete pointlessness of trying to keep a lawn green during a drought (they recover quickly afterwards anyway), a beautiful garden is possible without depending on large amounts of water from the mains.

Choosing plants carefully, collecting rainwater and using grey water from the house can all save on the amount you need to get from the taps, without reducing the enjoyment you get from the garden at all. In fact, a drought-friendly garden can be much less work to look after.

18 DRY GARDEN PLANTING

Choosing varieties of grass with lower watering needs can help your lawn survive a hot summer. These include varieties of ryegrass, fescues and meadow grass, and you can buy ready-prepared drought-tolerant grass mixes from garden centres.

In your pots and borders, choose plants that don't need a lot of watering. Look out for plants with small or hairy leaves, as well as shrubs with woody stalks.

Some vegetables, such as tomatoes and courgettes, need a lot of water. Choose less thirsty things to grow, such as spinach or peas, or put tomatoes in grow-bags, which help to retain water.

And don't forget to use a mulch around your plants. A layer of bark, gravel, fibrous compost or leaf mould helps to reduce evaporation from the soil and cut down on the need to apply water.

When you do need to give your plants a drink, use a watering can rather than a hosepipe, which can go through hundreds of litres in a few minutes.

Apply water to the roots of plants rather than the leaves, and time your watering carefully. The best times to water are early in the morning or after sundown. This gives the moisture time to soak right into the soil before the sun starts to evaporate it.

USE RAINWATER 19

Any gardening expert will tell you that flowers and vegetables actually prefer rainwater to tap water, so you'll be doing them a real favour if you fit a water butt to your drainpipe to capture some of the tens of thousands of litres of water that fall on your roof every year.

To collect even more water, several water butts can be joined together so they can fill up in turn as each one gets full.

You may be able to get a water butt free from your local council, and your local garden centre will stock a range of different sizes, including slim models that don't take up a lot of space.

Put your water butt on a stand so you can fit a watering can underneath the tap and, to keep children safe, make sure your water butt has a lid. Finally, to stop insects from breeding in it, float polystyrene balls or chips on the surface of the water.

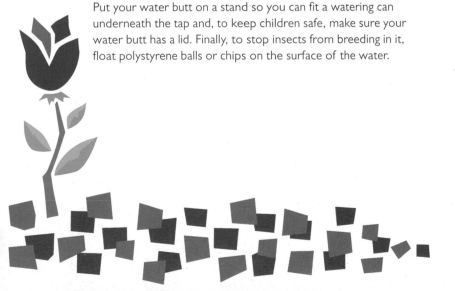

20 USE GREY WATER

Grey water from washing-up and showers can be recycled for use in the garden to save even more water.

Grey water is safe to use on the garden, although, to be extra careful, you should avoid using it on crops you plan to eat. And you should never store grey water in your water butt because this might encourage nasty bacteria.

It's easy to collect grey water in the shower. Simply place a bucket near your feet and it will catch the drips.

If you are still having the occasional bath, you can buy an adaptor kit to turn your hosepipe into a siphon to bring water down to the garden. If hosepipes are banned in your area, you can have fun explaining to anyone who asks that you aren't breaking the law – just watering your plants directly from the bath.

Washing-up water is also fine to use in the garden, provided you remove food scraps first. Pour your washing-up water through a colander into a bucket and it's ready to be applied to your thirsty plants outside.

WATER-SAVING HOME DESIGN

Think of a home with no plant life in or around it. No pots either side of the door, no hedge at the front, no trees lining the street, rooms inside with no houseplants and a garden without plants or shrubs, only bare stones and concrete. Not an inviting prospect, is it?

The desire to live alongside vegetation is second nature, but our towns and cities are slowly being stripped of their plants. Street trees are being blamed for subsidence and litter, and are being felled in their millions on the instructions of insurance companies and cautious officials. Front gardens are steadily being paved to create parking spaces, and back gardens are giving way to low-maintenance paving and decking.

This process of degreening our home towns doesn't just make them less pleasant, it makes them less healthy places to live and more susceptible to the effects of extreme weather, such as heatwaves and downpours.

Urban greenery has numerous benefits:
• Cleaning the air, removing toxic gases and tiny particulates.
• Cooling the climate, reducing the city 'heat island' effect.
• Providing a wildlife habitat for creatures such as insects and birds.
• Adding beauty and variety to the streetscape with the change in seasons.
• And – most importantly for this book – absorbing heavy rain to reduce the risk of flash floods.

Trees planted around a building can also reduce its energy needs by up to 25 per cent by providing shade in summer, while letting light through in winter.

By keeping the area around your home rich in plant life, you will gain more than just better water management.

21 | CREATE A LIVING ROOF

Living roofs (also called 'green' or 'brown' roofs) can be anything from a few centimetres of soil supporting lichens and mosses, to a fully planted up roof garden.

Given the huge area of unused roof space in towns and cities, converting just a small percentage to living roofs can make a big difference.

Benefits of living roofs

These go on and on, and don't only relate to saving water. Living roofs provide beauty and interest to people looking out over the roofscape, while absorbing carbon dioxide and air pollution.

But the most important benefit for water use is the effect of living roofs in reducing storm water runoff. Living roofs in Berlin absorb 75 per cent of the rain that falls on them for later drainage and evaporation, significantly reducing the amount of water that flows into sewers and reducing flood risks.

A living layer on a roof also protects waterproofing from ultraviolet light and the effects of frost, and increases the insulation value of the roof by up to 10 per cent.

While not all wildlife can easily climb up to a roof garden, living roofs do provide important habitat space for insects and birds – butterflies will fly up to twenty storeys above the ground to find nectar.

Where can I put a living roof?

Industrial living roofs are becoming common in many countries. The Ecover detergent factory in Belgium has a 10,000 square foot green roof covered in sedum, while the Ford Motor Company factory in Michigan has half a million square feet of green roofing, bringing a dramatic improvement in local water management.

Domestic living roofs are best suited to flat or shallow-pitched roofs on houses, garages or sheds. An extension to your home may well be a suitable spot, but make sure you get expert help to assess whether your roof can support the extra weight.

It is also important that the existing roof fabric is waterproof. A living roof will help keep your roof in good condition, but it must start out that way!

What type of living roof?

There are three types of living roof: extensive (shallow soil or gravel), semi-extensive (a greater depth of soil, supporting grasses or meadows) and intensive (deep soil, able to support larger plants).

Of these, extensive and semi-extensive roofs can be suitable for most homes, while intensive roofs are more specialised and best left for new buildings where they can be properly integrated into the plans.

What plants can grow on a roof?

Roofs are exposed to the elements and can dry out completely in summer, so hardier species do best on a living roof. Pick plants that grow in the wild on mountains, cliffs and deserts.

An extensive roof with very shallow soil is best planted with mosses or a green succulent called sedum, which resembles grass at a distance and is very tolerant to drought.

A deeper extensive or semi-extensive roof can support a wider range of plants. Drought-tolerant alpine plants and wild meadow flowers are suitable, although they are still likely to die back in a very hot summer.

Where do I start?

I can't emphasise enough the need for expert help in planning a living roof, because of the safety issues. However, lots of advice is available these days, and a simple extensive roof can bring great benefits and yet be a relatively easy DIY project.

The organisations listed on pages 124–128 can provide help and advice to get your living roof off the ground.

22 DON'T PAVE YOUR GARDEN

Green front gardens are becoming an endangered species, under threat from car parking and the fashion for no-maintenance, minimalist outside space.

This is a sad, but pertinent, example of seemingly harmless individual decisions combining to have a huge impact on our environment.

In London in recent years, we have lost an area of green space in front gardens the size of twenty-two Hyde Parks. If this was all paved over in one place, there would be an outcry, but because the damage happens in lots of small steps, it has gone almost unnoticed.

The same problem is developing in other countries, too, with around two thirds of urban front gardens now partly or wholly paved over.

The loss of front gardens in towns and cities causes a wide range of problems. Some of these aren't obvious – fewer green spaces will soak up less pollution, and safety for people walking on the street becomes an issue because of cars crossing the pavement. The loss of surfaces that can soak up heavy rain is one of the most serious effects, as this makes flash floods more likely.

One solution to this problem is obvious. Don't pave or concrete over your garden. If you have inherited a bare front yard from the previous owner of your house, rip up the paving and plant some easy-care shrubs instead, or simply replace the paving with gravel (not as good for wildlife or pollution, but a much more porous surface to deal with downpours).

In areas where a hard surface is essential, there are several options that increase the drainage potential of the ground.

Gravel is one alternative, as long as you don't put a waterproof membrane underneath it. Other options include paving slabs with integrated drainage holes or slabs made from recycled plastic or crushed glass that let water through. If you have to park your car on a driveway, consider using these materials in two strips to support the wheels, leaving an unpaved strip in between where hardy plants can grow.

SAVING ENERGY

Why save energy in the home?

Overall, around 30 per cent of our carbon dioxide emissions come from our homes, but we could reduce these by almost two thirds with the right energy-saving measures.

Most of these measures are easy to install and don't require daily effort. Examples include putting in better insulation, resetting your central heating controls, or getting more efficient appliances. Other measures can very quickly become part of your routine, and don't hurt your quality of life one bit – they may even make life easier!

Because heating up our homes uses so much energy, things that reduce the energy needed for heating are the best value. This doesn't mean shivering all winter. In fact, the energy-saving ideas in these chapters will make your home even cosier and reduce the cost of your bills at the same time.

Outside the home

Don't stop saving energy when you leave the house. Because reducing carbon dioxide emissions also saves money, all that's needed to get an energy-saving culture going at work is a bit of prompting from enthusiastic employees like you.

In your local streets and around your town as well, starting an energy-saving initiative can result in huge savings and a lot of fun for everyone involved.

INSULATE YOUR HOME

We'll start with insulation, because the best kind of heating is heating you don't have to do, thanks to having a home that keeps warmth inside rather than letting it out to heat up the street.

About three quarters of your heating will escape through your roof, windows, walls and doors if you have a poorly insulated, draughty house. Many of the measures you can take to cut down on this waste cost very little and pay back within a year or two in reduced bills.

Other improvements can take longer to make a profit, but remember, you will benefit from a cosier home throughout this time, and will be reducing your carbon emissions, too.

23 INSULATE YOUR ROOF

Around a third of heat losses occur through a poorly insulated roof. Insulating your loft space can reduce these losses dramatically.

Many different materials can be used, and at least 200mm (preferably more than 300mm) is needed.

The cheapest material is glass fibre insulation, which comes in huge rolls and is very easy to install. Polystyrene is also commonly used where flat boards are preferred (or it can be used in expanded form to fill cavity walls). However, these materials contain lots of chemicals and take a lot of energy to produce. Glass fibres can also be a health hazard, so protection must be worn when installing this kind of insulation.

There are many natural materials that are suitable for use in insulation, many of which use very little energy in production, don't contain so many chemicals, and can be recycled after use.

Look out for insulation made from:
• Sheep's wool (known as Therma Fleece)
• Recycled newspaper (known as Warmcel)
• Hemp or flax fibres (brands include Isonat and Flax 100)
• Wood fibre boards (pressed waste from the timber industry)

An uninsulated roof can contribute to up to one third of heat loss from your home

The same amount of heat can be lost through uninsulated walls

A further 20 per cent can be lost through single-glazed windows

24 INSULATE YOUR WALLS

Most brick homes built in the last one hundred years have cavity walls, with two layers of bricks and a small cavity in between. Filling this cavity with insulating material is simple and cheap and can pay back in reduced bills within two years. To fill the cavity, engineers simply drill a hole in the wall and pump in material. It can take just a few hours, but the benefits last a lifetime and you can usually get grants to help with the cost of cavity-wall insulation.

Older houses that have solid walls can be more complicated to insulate, but can save even more money on bills. Payback times are longer, because the process is more labour-intensive and requires skilled work, but there are some grants available.

External insulation
This involves cladding your home with a thin layer of insulating material, followed by new exterior facing. This is the best option as it also maintains the 'thermal mass' (see page 66) of your home.

Internal insulation
This involves putting a similar thin layer of material straight onto interior walls or building a wooden frame which is filled with insulating material and covered in new plasterboard.

Internal insulation will reduce heat loss very effectively, but it also reduces thermal mass, and cuts down on the interior space by a small amount. However, it is cheaper than external insulation and is most suitable for a flat, as only the walls that lead outside need to be covered.

IMPROVE WINDOWS 25

Windows that are draughty and have single-glazing can be responsible for about a fifth of the heat escaping from your home.

Draught-proofing

Putting draught-proofing strips around windows is a simple task that you can do yourself in a few hours. These strips are generally made of self-adhesive foam that squashes and creates a seal when the window is shut.

Don't forget to deal with the draughts from doors. Front doors should have an insulating brush over the letter box, and all exterior doors should have a brush strip fitted across the bottom.

Double-glazing

Double-glazing cuts heat losses through windows by about half. Even if you live in a conservation area you should still be able to get replacement double-glazing to match – most kinds of wooden window frames come in double-glazed versions nowadays.

Windows now have energy efficiency labels in the same style as those you find on fridges and freezers. The British Fenestration Rating Council can help you search for highly rated windows in the style you need, and help you find a company in your area to install them.

NATURAL HEATING

You can use the power of the sun to heat your home without high-tech gadgets simply by utilising well-placed windows and a property of your home known as 'thermal mass'. A home with high thermal mass not only stays warmer in winter, but also cooler in summer, bringing double the benefits.

Creating barriers between the inside and outside of your house doesn't only have to be done with insulation. A porch can be attractive, as well as an effective 'heat lock', keeping cold air in its place.

The right kind of glazing in the right place can capture the heat of the sun and bring it into your home. If you can't move all your windows around (most of us can't!), then a conservatory may be the answer. These are useful and attractive, and can add value to your home as well.

26 MAKE USE OF THERMAL MASS

We have all walked into a castle or cathedral on a hot day and noticed how cool it stays inside. The reason is that these buildings have a high 'thermal mass', i.e. the large amount of stone in the walls takes a long time to heat up and cool down, and therefore helps to regulate the temperature inside.

In the case of the cool cathedral, the thick walls, having cooled down overnight, continue to absorb heat right through that hot summer's day, keeping the temperature down.

Few of us visit castles in the dark, but if we did we would also find that the walls stay warm through the night, keeping the temperature inside higher. You might have spotted how a stone wall remains warm long after the sun has set – again this is because of the high thermal mass of this material.

In a badly insulated home, the opposite processes occur. On a hot day, the walls absorb heat rapidly and won't help to cool the room inside; and on a cold night, heat will pass quickly through the walls, making everything inside chilly as well.

Increasing your home's thermal mass

If you are building a new house, you should always think about thermal mass when deciding which material to build with.

But it's not just the walls that contribute. Everything in the home goes towards its total thermal mass, so using materials such as stone for a fireplace or flooring can make a real difference to the thermal mass of existing homes.

If you are adding an extension to your home, use materials with high thermal mass properties and they will help regulate the temperature throughout your home.

Materials with high thermal mass include:
• Brick (for walls and floors)
• Concrete (but has a high carbon footprint)
• Stone (best used for floors)
• Water (which is why central heating radiators are a good idea all year round)

The ability of materials to absorb heat is also increased when they are dark in colour. Try placing dark-coloured stone on the floor of your conservatory or kitchen, or paint walls a darker colour opposite south-facing windows to capture the heat from sunlight that falls on them.

Where you place your insulation also makes a difference. Cavity-wall insulation removes the outer layer of bricks from the thermal mass equation, and internal insulation, while excellent for keeping warm in winter, reduces the thermal mass capacity of the walls considerably and will make you warmer in summer as well.

GLAZING FOR 'SOLAR GAIN' 27

The best way to capture the heat from the sun is to have windows in the correct places around the home.

An efficient solar house (in the northern hemisphere at least) has more windows in south-facing walls, and fewer, smaller windows – or none at all – to the north. This makes sure that morning heat is captured to the fullest in winter, and prevents heat escaping on the colder side of the house.

East- and west-facing windows are more problematic. It is relatively easy to shade a south-facing window with an awning or over-hanging eave, but in summer, west-facing windows can cause over-heating, as they will receive a direct hit from the lower afternoon sun, which makes them difficult to shade.

A clever way to avoid excessive solar gain in summer is to place deciduous trees and shrubs outside south- and west-facing windows. In winter, when solar heat is needed, they will be bare and let the light through. In summer, they will be covered in leaves and provide cooling shade instead.

Double-glazing

Make sure you use double-glazing or you will find that heat losses through single-glazing in the winter will outdo any gains from solar heating. Look out for double-glazing with a low-emissivity coating on the inside pane. This coating lets sunlight in to warm the room, but reflects infrared light coming from inside the room, reducing the heat that escapes back out again.

28 A PORCH OR CONSERVATORY

An exposed front door that opens into your hallway lets in a blast of cold air each time it is opened, seriously reducing energy efficiency.

A porch that adds an extra door between you and the outside helps to reduce this effect by acting as a heat lock. Porches are attractive, too, and provide a useful storage space for muddy shoes and umbrellas. You can even use your porch as a mini greenhouse for kitchen herbs and tomatoes.

A conservatory can be a surprisingly effective way to make the most of solar gain in winter. The large area of glazing effectively preheats the air before it comes into the house, and the conservatory also acts as a layer of external insulation. With effective ventilation and high thermal mass materials, a conservatory can also help to cool your home in summer.

Apart from the potential energy-saving benefits, a conservatory is a useful addition to any home, providing a sunny space to grow plants and to sit on warm days, and it will even add to the value of your home if it is well built.

It is important to think carefully about the design, materials and orientation of your conservatory if you decide to build one. A badly designed conservatory that is open to the rest of the house all year round could actually double your heating bills!

Bear these tips in mind if you plan to get a conservatory:

• Keep it separate from the rest of the house. Build one that you enter via existing doors, and make sure you can seal it away on cold days in winter.

• Put your conservatory on the south side of the house, or the south-east if this is not possible. Design it so that the main glazing is on the south-facing walls.

• Use high thermal mass materials, such as brick and stone, for the walls and floor. The heat from the sunlight that reaches these materials will be stored for release later, increasing the heating effect of the conservatory.

• Don't heat your conservatory with any artificial means – leave it to the sun alone.

• Fit a window vent to pull the warmed air into the house. Remember to close the vent on very cold days.

• Make sure the conservatory can be ventilated high up in the windows or through its roof. This will allow you to use it to help ventilate the house in summer, and prevent it getting too hot.

LOW-CARBON HEATING

Solar gain and insulation alone are unlikely to be able to provide all your heating needs, especially in winter. So, what is the best way of providing heating in the home? There are lots of different options out there and it can be confusing.

On the whole, systems that use radiators or pipes filled with hot water are the best at providing gentle, long-lasting heat. Water has a very high thermal mass, so heat is maintained in the system long after the boiler is turned off, and pipes and radiators don't encourage dust to move around the room in the way some air-heating systems do.

There are lots of ways of heating the water in your radiators, and as more sustainable options become available, your system can be adjusted to run off these, too. To help you decide the best option for now, this chapter will go through the benefits and problems of a few fuels.

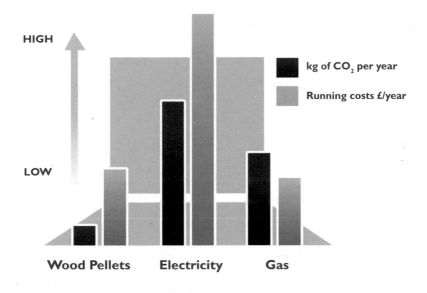

HIGH

LOW

kg of CO_2 per year

Running costs £/year

Wood Pellets   **Electricity**   **Gas**

29 HOW TO HEAT

ELECTRICITY

Electric heating is one of the least efficient ways of turning fuel into heat. With electricity generated from gas, only about a third of the energy in the gas eventually reaches your home to be used for heating, whereas an efficient gas boiler can turn more than 90 per cent of the energy in the fuel into heat.

Electricity is far better used for lighting and powering appliances rather than producing heat, so avoid electric heaters.

SOLAR

A lot of the sun's energy already reaches us in the form of heat, so capturing this for the home is relatively simple. As well as the 'passive' measures in the chapter on solar gain, you can also use active solar power to heat water for your home. In new houses, solar panels can also be used to heat water for underfloor heating. However, if you are not building from scratch, the best way of using solar power is to use it to heat water for your showers and hot taps.

How does solar hot-water heating work?

There are two main kinds of solar water-heating panels. One uses a dark-coloured flat plate behind glass to capture heat and transfer it to a fluid in copper pipes behind the plate. A more efficient system has multiple copper collection plates held in vacuum tubes. This reduces the heat lost through contact with the rest of the system and ensures more of the sun's energy is transferred to the circulating fluid.

How much hot water can I get?

In summer, a solar hot-water system is easily able to create hot water at sixty degrees centigrade without any extra heating. In winter, the system will still be able to preheat the water and save energy, but back-up heating from a gas or wood-fired boiler will be needed to make properly hot water. Overall, you can get more than half your hot water needs from a reasonably sized system (between 2–4 square metres), even in northern Europe.

Because solar water heating makes use of renewable energy, grants are usually available to help with the cost of installation.

HEAT PUMPS

Heat pumps work in the same way as your fridge, but in reverse. Instead of using a liquid to transfer heat out of a box, they capture heat from outside the home, concentrate it and then transfer it to the water used in your heating system.

Heat pumps are a very efficient way of using electricity for heating, because for every 1kWh of electricity, 3kWh of heat is generated. This makes their overall running costs about the same as for gas, but with lower emissions. The exact saving depends on where the electricity is coming from; the most eco-friendly combination is a heat pump running off electricity from renewable sources.

Ground-source heat pumps work most efficiently, as the fluid-filled pipes are either sunk into a borehole or buried in trenches a metre or so under the ground. Below this depth, the temperature remains about the same all year round. Air-source heat pumps carry out the same process using the heat from the air outside, but don't work well in very cold weather.

GAS

The best condensing boilers can reach 90 per cent efficiency. Condensing boilers are more efficient than older boilers, because they are designed to capture more of the heat from the exhaust gases.

If your boiler is more than about fifteen years old, you will see up to 35 per cent savings on your gas bill by replacing it with a new A-rated boiler (with similar reductions in carbon dioxide emissions), making this a very worthwhile investment.

WOOD

High-performance wood-pellet boilers, which convert fuel to heat at around 80 per cent efficiency, are now available.

Because the fuel they use has been grown recently (capturing carbon from the atmosphere instead of from fossil fuels), the effective carbon dioxide emissions from this form of heating are very low. The pellets are made with waste from the timber industry rather than from unsustainable logging, and these heating systems are also relatively economical to run.

Unlike traditional wood stoves, which need refuelling regularly, modern machines have a fuel store within the burner, which only needs topping up every few days.

BE IN CONTROL 30

We all know that turning your thermostat down by one degree can shave 10 per cent off your heating bill, but with better timed heating and more sensitive controls, you can save even more. Only using heating when you need it and controlling the temperature more precisely can take a big chunk out of your contribution to climate change.

If your heating system has a simple on/off switch, or runs off just one basic thermostat, a more sophisticated control panel could help.

Modern digital controls enable you to set the desired temperature to within half a degree, and have your heating on at different times of the day on different days of the week. They can also enable water- and space-heating times to be set separately.

New control panels can be fitted even if you are not replacing the boiler. Adding individual thermostat controls to radiators means you can reduce the heat delivered to rooms when they are not in use. Don't fit a radiator thermostat in the room where your main heating thermostat is because they might confuse each other and other rooms may get very hot!

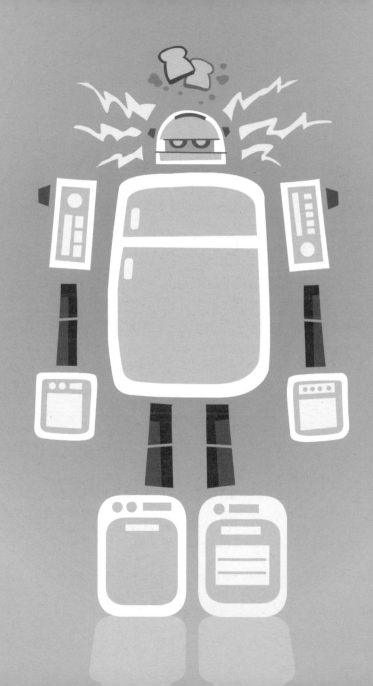

SAVING ELECTRICITY

As we reduce the energy we use for heating, the proportion of our carbon footprint coming from the electricity we use to power lights, appliances and gadgets around the home goes up. In newer houses with good insulation, electricity is responsible for around a third of all carbon emissions.

The next most important item on our energy-saving list is therefore saving electricity.

In fact, the total amount of electricity used in homes has doubled since the 1970s and is still going up rapidly as we get more pieces of equipment and use them more often.

Many appliances bring real benefits and help us cope with busy lives, so I'm not going to suggest giving them all up (OK, maybe some of the more pointless ones), but it's easy to let machines we use only occasionally become 'vampire electronics' by leaving them on standby, where they suck power from the grid all the time, while doing nothing useful at all.

By switching to greener lights and electronics, and by turning them off properly and making use of their built-in energy-saving features, you can cut your electricity bill without cutting back on creature comforts.

3 | WATCH YOUR USAGE

In any electricity-saving drive, it helps to know what the results are if you're going to keep the whole family motivated.

Your electricity bill is a good guide to how much energy you are using up. The amount of energy is shown on your bills in kilowatt-hours (kWh). A kWh is a unit of electrical energy equivalent to boiling the average 1 kilowatt kettle for about an hour. Generating this energy is responsible for releasing carbon dioxide into the atmosphere and contributing to climate change.

The amount depends on how the electricity is generated, and the mixture of methods used in the UK means that each kWh from the national grid represents about 600 grams of carbon dioxide emissions from power stations.

So, if your electricity bill says in three months you have used 1,000kWh, this means your electricity consumption generated 600kg of carbon dioxide – this is about the same as driving a family car for about 2,500 miles.

GREEN YOUR LIGHTS 32

We are getting more sophisticated in the way we light our homes, rejecting harsh overhead lights in favour of a range of small lamps and integrated fittings in cupboards and alcoves. The average home now has twenty-three light bulbs, with an increase of three more expected within fifteen years.

Too many of these light fittings are still using old-fashioned incandescent bulbs. Low-energy compact fluorescent bulbs now come in sizes to fit almost every kind of lamp, so there's no reason to carry on using bulbs that use up four times the energy and last only a twelfth as long.

If every house replaced just three bulbs with energy-saving models, we would save enough electricity to power all our street lamps, and if we replaced all twenty-three bulbs, the savings would be colossal.

It can still be hard to find a proper range of efficient light bulbs in high-street shops, but online eco-shops tend to stock all the different sizes. Why not visit the websites at the end of this book and order bulbs to fit the lamps you haven't changed over?

33 MONITOR ELECTRICITY

Electricity bills only arrive about four times a year – not very helpful if you are trying to measure your savings as you put in eco-friendly improvements week by week.

A great way to keep an eye on electrical waste and monitor the success of your energy-saving efforts is to get a clever device that reads your meter and shows your energy use – in real time – on a portable display you can keep anywhere in the house.

Two products currently on the market are the Electrisave and the Efergy monitors. The readouts on these machines show exactly what your meter is doing, minute by minute, so you can switch on the TV or a lamp and it will show you instantly how many kWh you are going through. If you find it hard to think in kWh, then don't worry, you can also get the display to show carbon dioxide emissions or even the cost.

These devices are an excellent way to prompt you and your family into adopting better energy-saving habits. A quick check of the readout before bed will also tell you if you have left anything on.

When people have solar panels and wind turbines fitted to their homes, they also get a meter monitor like this to help them see what they are generating as well as what they are using up. As a result, they use up to 25 per cent less electricity, simply because they are prompted to cut down on waste.

34 CLEAN MACHINES

Choosing the best household consumer durables and using the machines you own wisely can make a massive difference to the amount of electricity your home consumes.

White goods (e.g. fridges and washing machines)

Between them, white goods are currently responsible for about 40 per cent of our household electricity consumption.

Since the introduction of energy labels in showrooms, choosing the best appliances has become a lot easier, and in most shops it is hard to find a model that isn't rated A or A+. Overall, the total electricity used by these appliances is going down by about 2 per cent a year.

Follow the tips in the kitchen section to help you make the most of their energy-efficient features.

Electronics

How many items from this list of gadgets did you have at home when you were little?
- Video recorder
- DVD player
- Answering machine
- Games console
- Set-top box
- Mobile phone

Because of my dad's obsession with watching cricket matches during the night, we got a video recorder in about 1982, which was then a huge novelty. Nowadays, most households have all of these items as a matter of course.

Unlike washing machines, newer electronics use up more electricity than their predecessors. Extra functions and higher specifications can make the latest models very energy-hungry – a new flat screen plasma TV can take twice the energy to run compared with the box it replaces.

All this adds up to trouble. By 2010, electronics will have overtaken white goods to become the biggest consumer of electricity in the home – 45 per cent of all consumption will be from entertainment, computers and gadgets.

How can we be energy-saving with our electronics without cutting down on the fun we get out of them? Try these tips:

• Choose the lowest energy electronics products you can find. These machines don't have to show energy labels, but the power consumption should be written on the box. You can find wide variations between very similar products, so it is well worth hunting out this information.

• If you absolutely have to have a massive TV, get a projection model, not a plasma screen. Power consumption goes up quickly with screen size for plasmas, but a TV that projects its picture onto the wall can be as big as you like for the same power consumption. If screen size is less important, the most energy-efficient TVs are flat-screen LCD models.

• Get machines that combine different functions. I have recently replaced a very old VCR with one that plays and records DVDs as well. Many TVs now come with an integrated digital set-top box.

35 VANQUISHING VAMPIRES

Possibly the most wasteful thing about the rise in electronics in the home is the number of these machines that don't come with a proper off switch. Televisions, VCRs and DVD players are all notorious for being controlled by a remote control that can only put them into standby mode.

Some of the worst machines for using up standby power are, surprisingly, games consoles. If not shut down, these machines remain in 'idle' mode indefinitely, consuming virtually the same amount of power used while playing a game.

Other things that 'should know better' in environmental terms include gadgets that need charging up with an external adaptor. Rechargeable machines could help the environment because they don't need lots of replacement batteries. But chargers are very easy to leave plugged in, so can waste energy on a massive scale – more than a billion new chargers are produced every year worldwide.

Try these tips to tame the vampire electronics in your house:

• Turn things off, either at the plug or with their manual off switch, whenever you can.

• If your machines don't have off buttons, get extension leads with individual switches for each plug. These make it much easier to turn off individual items rather than leave everything else on standby just because one of the sockets is being used.

• Look for appliances which have low standby power consumption. This is usually shown on the box, so aim for standby consumption of less than one watt.

• All chargers should be unplugged when they are not being used. To avoid having to charge your phone overnight, get into the habit of plugging it in as soon as you get home in the evening. By the time you go to bed, it will be fully charged and safe to unplug ready for the morning.

36 THE AGE OF THE COMPUTER

Setting up a home office can have a big effect on the electricity consumption of your home. Per year, a single computer, monitor and printer can add more than £100 to your annual electricity costs if you aren't careful about using them.

Computers use up standby power, but they are also very likely to be left fully powered up for long periods. Computers are also more likely than other electronics to be left on overnight. Beware of seemingly worthy programmes that use your computer's 'spare' capacity to do things, such as search for alien messages or analyse DNA sequences. Running these programs will mean your computer uses almost as much power as when you are working – and you will end up paying for it.

Combined with the fact that it is so easy to leave printers, scanners and other accessories switched on, home computing is full of little ways to waste energy, so here's how to cut down.

Desktop or laptop?

Laptops use on average 70 per cent less energy than a desktop computer, so if you can, get a laptop instead. You will save space as well as plenty of carbon dioxide and a big chunk of money.

The most energy-efficient laptops have Energy Star labels awarded by the US government, so look out for these models. Also beware that computers contain a lot of toxins and metals and absolutely must be recycled at the end of their lives. Most recycling centres will now take electronic waste for specialist treatment, so never put a broken computer or laptop in the bin.

Choosing a computer

Like other electronics, higher-spec computers tend to use more energy, as their processors and monitors get faster and larger. But there is a lot of variation, even between quite similar products, so check the small print before you buy, and don't buy a machine with tons more processing power than you need.

Remember to look at the power consumption in standby and sleep mode, as well as the in-use power requirements of your computer. Because sleep and standby can be used for long periods, even a small difference can add up to large amounts of waste.

SLEEP, STANDBY OR OFF?

Computers use different amounts of power depending on how hard they are working. Wander off with programs running and they will still use up most of the power they consumed while you were working on them. Screensavers may look pretty, but they are definitely not a way of saving energy. This is known as idle mode and is a very wasteful state in which to leave a computer.

Sleep mode

You can very easily set up your computer to go into sleep mode if you stop using the keyboard or mouse for a certain length of time. Choose a sensible delay, such as five or ten minutes, so that, after this time, your computer will shut down most of its power-using functions but remain ready to spring into life once you return.

Standby

When you shut down your computer and all its lights go out, you might think it isn't any energy at all. But no, even in standby mode, some components inside stay active, using up to ten watts of power in some machines. Used continuously, this adds up to £8 or £9 in electricity costs in a year. Not a lot, but much better donated to your favourite charity than wasted on a computer doing nothing.

Off

The only way to be certain that your computer really is properly off is to switch off the power at the plug. Incidentally, this is also the most effective firewall – a powered-down computer is 100 per cent impregnable to hackers and other nasties.

Monitors

Most office computers now have flat LCD screens, but for home computing, it's more likely you will still be using a chunky cathode ray tube (CRT) monitor.

As well as taking up space, CRT monitors also give off a lot of heat and can use around five times more energy than an LCD monitor. Upgrading your screen can save you nearly £30 per year in running costs, but remember that power consumption goes up quickly as screens get bigger, so don't get one that's bigger than you need. Again, look for Energy Star models for the best savings.

Printers

Printers use up even more idle power than computers, rarely have sleep modes, and the off button on top may not shut them down completely either. Again, switching off at the plug is the answer. If your sockets are hard to reach, a great investment is a multi-plug adaptor with individual switches, so you can label the plugs and switch off the printer (and other peripherals, such as scanners) except when you actually need it.

USING COMPUTERS FOR GOOD

Home computers aren't all bad news for the planet. You can make effective use of the internet to save energy, too:

• Get things delivered by mail instead of travelling to shops.
• Seek out greener products online.
• Make use of downloading services for music and films to cut out the energy used to manufacture CDs and DVDs.

GENERATING ELECTRICITY

Almost any kind of energy can be captured and turned into electricity. At the moment, most electricity is made by big power stations burning fossil fuels, such as coal, oil and gas, which contain the energy captured from the sun by plants millions of years ago.

Fossil fuels aren't a renewable source of energy. In fact, they are running out fast. And even if they could last much longer, we couldn't use them all up without doing huge amounts of damage to the world's climate with all that carbon dioxide.

Nuclear power is another dirty way of generating electricity. Like fossil energy, nuclear energy comes from a non-renewable resource: scarce uranium ore, which needs to be mined and processed before it can be used in power stations. Because of the huge scale of nuclear power stations and the numerous safety problems, nuclear energy is expensive to harness and creates large amounts of dangerous waste, which we still don't know how to deal with.

Clean, renewable energy from the wind, sun and tides won't run out and doesn't cause pollution, and there are a range of different technologies – both new and old – which are being used to harness this green energy. Each has its own advantages and problems, and all of them will have to be used in combination to provide electricity in the future.

Energy from the sun

We have already seen how warmth from the sun can be used to provide hot water. Solar photovoltaic cells are made from semiconductors that turn light energy directly into electricity. They have no moving parts, so are very easy to maintain and can be integrated into buildings to provide power on site.

In the sunniest areas of the world, such as California, large solar arrays are being used to generate as much electricity as a conventional power station, and in areas where grid connection is difficult, solar power is often the cheapest way of providing electricity.

There are lots of advantages to solar electricity, but the obvious drawback is that it can only be generated during the day.

Wind energy

Wind energy is one of the oldest forms of renewable energy that humans have used – sailing boats have been powered by the wind for thousands of years, and windmills have been common for centuries.

Wind turbines that generate electricity range from tiny 1kW machines on homes and caravans to huge 3MW turbines out at sea. Wind power is a fast-growing industry. More than 25,000MW of capacity is installed around the world, but this is only a fraction of the potential of this renewable resource.

Like solar, the amount of wind power varies with the weather, but as wind speed can be predicted by weather forecasts, it is a very useful contribution to the electricity mix.

Water energy

This comes in many forms, including the energy held in water that wants to flow downhill and the energy of fast-flowing water and ocean waves.

Hydroelectric power was developed in the twentieth century, when many of the world's rivers were dammed to provide power. The giant Hoover Dam in the USA was built in the 1930s across the Colorado river, and can generate 2,000MW of power, providing power every year to homes in Arizona, Nevada and California.

The great advantage of hydroelectric power is that it can be turned on or off, depending on demand. However, large dams cause a lot of environmental damage, so building more of these isn't a good idea.

Tidal power makes use of the rise and fall of tides under the gravity of the sun and moon. Power is generated by trapping water at high tide behind a barrage or in lagoons. The water can then be allowed to flow through turbines at low tide to generate electricity.

Fast-flowing tidal currents can also be harnessed with underwater turbines, although this technology is still at quite an early stage of development.

Waves are a free, endless source of energy created by winds blowing across the surface of the ocean. It is the most concentrated form of renewable energy, but it is largely untapped.

Various kinds of wave turbines can harness this energy to produce electricity either on the shoreline, where the waves break, or in deeper water offshore. Onshore wave turbines are ideal for a harbour wall where waves break. They work by getting the waves to push air into narrow chambers to drive turbines.

Other wave devices sit on the surface of the ocean and bend with the passing waves, creating electricity using hydraulics from the motion of their joints (see picture below). The first large-scale wave power stations are being built using this technology in Portugal and Scotland.

Biomass
Renewable electricity can be generated by burning organic material, such as wood, straw or animal waste. The process is low-carbon, because the materials burned have grown recently, so the carbon dioxide they release doesn't add to the overall level of carbon in the atmosphere.

Biomass is best used when the material being burned is really a waste product that would otherwise go to landfill. Since 1998, a power station fired with poultry litter from nearby farms has been generating electricity for the town of Thetford in Norfolk, which has proved a very useful addition to local electricity supplies. However, turning over large areas of farmland to growing crops specifically for burning in power stations isn't a good idea, because of the effect on food production. Burning materials that could be used for other purposes, or recycled into new products is also a very poor use of biomass and, overall, a waste of energy.

One advantage of the smaller scale of biomass power is that it can be located in places where the waste heat can be used to provide heating for nearby homes and businesses. In big power stations, which are usually located in the middle of nowhere, this heat is simply released into the atmosphere where it really does go to waste. Smaller, local combined-heat-and-power plants are a much more efficient use of fuel and much better for the planet.

At home, biomass is most suitable for providing heat via a wood-pellet boiler or stove – see the heating section for more details.

37 BUY GREENER ELECTRICITY

You can help encourage investment in new wind, solar, wave and tidal energy by opting to pay for greener electricity through your electricity company.

The best policy, as used in Germany and many other countries, is one that guarantees higher payments on the electricity market for people and companies who generate renewable energy. By ensuring a good price was paid for green electricity, the government helped to make wind turbines and solar panels an attractive investment, massively increasing capacity in a few years.

In the UK, the government adopted a different policy called the Renewables Obligation. This meant that electricity companies had to buy a certain proportion of their electricity from renewable generators. Because of this, the companies could create 'green' tariffs simply by reallocating the renewable energy they had to buy anyway to these customers, creating no extra capacity at all.

Good greener tariffs
Don't despair: you can still make a difference by opting for a green tariff if you make sure your electricity company actually converts some of your money into new windmills, wave farms and solar panels.

Ethical Consumer magazine gives each a simple rating to help you make up your mind (www.ethiscore.com). In 2007, the National Consumer Council also produced a report called 'Reality or Rhetoric', which rated the different tariffs available to customers (www.ncc.org.uk).

GENERATE YOUR OWN 38

Both large- and small-scale green energy projects will be needed to provide our electricity needs in the future. By generating renewable energy at home, you can reduce your bills, help support a new industry and cut your carbon emissions.

Different technologies are suitable for different houses, and the best option for you will depend on where you live, as well as the structure of your home, so do get expert advice before purchasing any kit. Renewable energy systems that have been used successfully in homes include solar panels, rooftop wind turbines, wood-pellet boilers and ground- and air-source heat pumps.

Grants to help install renewable energy at home are also available from the government, although they are usually in short supply.

In the UK, the Energy Saving Trust and the Low Carbon Buildings Programme can help find expert advice and any available grants. Visit their websites for more information:
www.energysavingtrust.org.uk
www.lowcarbonbuildings.org.uk

SAVING ENERGY IN THE KITCHEN

With 15 per cent of electricity being used for cooking, and some of the most energy-hungry appliances living in the kitchen, this is a great place to look for energy-saving ideas.

There are plenty of ways to save energy while cooking, and there is a lot more you can do to reduce the impact of appliances, such as fridges and freezers, than simply buying A-rated machines.

39 SAVE ENERGY AS YOU COOK

Cooking is responsible for 6–8 per cent of our total energy consumption in the home, with ovens and hobs accounting for about half of this, and kettles up to a third.

The rest is from toasters, microwaves and the wide range of choppers, mixers, blenders, grills and toasting machines that we all accumulate, seemingly whether we like it or not. I don't know how many people still use the smoothie makers they got for Christmas last year?

With cooking taking most of the energy we consume in the kitchen, there are some simple measures to keep waste to a minimum.

• Use the right size pan for your food, and the right size ring or burner for the pan.

• Put a lid on saucepans for faster cooking, which also uses less energy. With a tight-fitting lid, you can turn the heat right down.

• Don't overfill the pan with water; just enough to cover the food is fine, especially if you use a lid to catch evaporation.

• Use the kettle to boil water for cooking, rather than heating it up in the pan – the only time you shouldn't do this is if you are boiling cold eggs from the fridge, as these may crack from the shock of having boiling water poured over them.

• Only fill the kettle with as much water as you need right then.

• Pasta will cook without being held at a rolling boil. Add pasta to hot water, bring it back to the boil and then turn off the heat and put on a tight-fitting lid. After the normal cooking time, you should find the pasta is ready to eat.

40 MICROWAVE MAKEOVER

Almost all of us now have a microwave oven in the kitchen. These can use much less energy because, unlike in a traditional oven, the waves only heat up the food – not the whole appliance.

Microwaves are great for quick, low-energy heating up of leftovers and are ideal for heating up a cup of milk for hot chocolate at bed time. You can also 'steam' vegetables (use a small amount of water and a tight-fitting lid). Steamed puddings can take hours to boil in a pan, but only need minutes in the microwave.

A really good use of your microwave is for baked potatoes. A spell in the microwave cuts down drastically on the cooking time needed, but that final twenty minutes in the oven is essential if you want delicious, crispy skins.

However, microwave cooking can have its pitfalls if you end up buying a lot of convenience foods. That ready-meal from the supermarket may only spend five minutes in your microwave, but it represents a lot of energy in total.

The typical ready-meal is made from ingredients brought from all over the world, cooked in a factory, then frozen, wrapped up in lots of packaging (which also needs to be transported to the factory), then taken by lorry to the supermarket before, finally, taken home in your car, on your bike or on the bus for a quick reheating in the microwave.

In comparison, a meal cooked at home from fresh, local vegetables (which will have used only minimal fossil fuels in growing and transporting) will consume less energy overall even if you cook it for an hour in the oven.

And of course, fresh food can also be much healthier than salty, processed ready-meals.

41 A-RATED APPLIANCES

Energy ratings are based on the most eco-friendly usage patterns, which aren't what we always manage in our daily lives. It's a bit like the mileage figures for new cars. Cars are tested on a treadmill that simulates really sensible driving, so it's actually quite difficult to achieve the same mileage in the real world.

'Cold' appliances

Fridges and freezers, in particular, can turn into energy hogs if they aren't maintained and used carefully. Try these tips to keep your appliances as planet-friendly as their eco-labels suggest:

• Buy a fridge or freezer no bigger than you need.
• Keep them away from your boiler and oven, or they will have to work harder than necessary. The ideal place for a freezer is in an unheated garage or cellar.
• Keep freezers as full as possible, but empty the fridge of old items regularly.
• Never put warm items into a fridge or freezer.
• Defrost fridges and freezers regularly, as a build-up of ice makes them run less efficiently.
• Remove dust from the condenser coils at the back of your fridge, so it can quickly radiate the heat away.
• Check your door seals. A piece of paper should stay stuck in the seal even if tugged gently. If you have a fridge or freezer that regularly ices up, the seals may well be at fault – and costing you a fortune, too.

'Wet' appliances

The energy rating for your washing machine is based on the most efficient washing cycle, usually labelled 'eco' on the dial.

This programme may take a bit longer to finish, but waiting an extra twenty minutes for your washing is rarely a problem. Use the eco-cycle whenever you can and you'll make the most of your green machine.

The test results are also based on washing a full load. This literally means filling the machine up to the top. While your instinct may be to leave the clothes with room to breathe, it's far better to pile them in right to the top. They will lose volume as soon as they get wet anyway.

EMBODIED ENERGY

Almost everything has an energy cost. When trying to live a greener life, we tend to worry most about activities that use up fuel directly, such as travelling or heating. However, the things we buy also contribute indirectly to climate change.

The climate impact of our purchases depends on three things:
• The energy used to make them (which comes from mining and processing raw materials and manufacturing);
• The energy used to transport them from factory to shop;
• And the energy used in their working life.

For a car, the energy used to make it is only a small proportion of its overall carbon footprint, which is mainly made up of the fuel burned while the car is being driven. Similarly, much more energy is used up washing and drying a piece of clothing in its lifetime than is used up in manufacturing it.

But for other items, such as furniture, toys, building materials and books, the embodied energy is the largest source of carbon emissions. Many goods are traded internationally, and embodied carbon emissions don't appear on the balance sheet of the countries where goods end up. If they did, places where the majority of manufactured goods are bought would have much higher recorded emissions, while countries such as China – where a lot of things are made – would have lower emissions.

So, even though it won't save money on your bills or help the government meet its targets, reducing the embodied energy in the things you buy is another important way to save energy. There are lots of ways of doing this, from making things last longer, to buying recycled goods, to choosing things made from materials with low embodied energy.

42 REPAIR, REUSE AND RECYCLE

Avoiding things made from newly minted raw materials helps to cut emissions. The three Rs is the simplest way of remembering the basic principles.

Repair
Make the things you own last longer by getting them repaired. Re-heeling your shoes regularly to keep them in good condition doesn't only save resources, but also all the time it takes to break them in. I usually find a catastrophic heel break occurs just after I have finally got my favourite pair of shoes to be comfortable.

Reuse
For high-quality second-hand furniture, look out for twentieth-century design classics at markets and auctions. Vintage and second-hand clothes can be a stylish bargain, as well as low in energy, and the number of designers making clothing and shoes from recycled materials is growing by the day.

Recycle
When items really are past their use-by date, make sure you put them in the recycling bin, so that the raw materials can be used again without the energy-intensive process of making them from scratch. More than 300kg of carbon dioxide is saved for every ton of recycled glass used instead of new material.

BUY LOW-
CARB GOODS

43

The embodied energy of most goods isn't listed on the label, and can be hard to work out. But there are ways to be careful about the amount of energy in the things you buy without taking your calculator and a pile of reference books to the shops.

Some general tips:

• Recycled or reused materials have a fraction of the embodied energy of new materials.

• Products designed to last a long time and which are used over and over again consume much less energy than so-called disposable products.

• Locally made goods are best, reducing the energy needed to get them to the shops. Reducing transport emissions has the most impact for heavy items, such as liquids.

• Metal items have a very high embodied energy, because of the high temperatures needed to extract metal from ores and then forge it into plates, sheets and beams.

• Timber has a much lower embodied energy, but watch out for wood that is not grown and harvested sustainably.

The embodied energy in the food we eat also leads to a lot of carbon emissions. See the section on saving water for more on reducing the environmental impact of our diets.

SAVING ENERGY AT WORK

Workplaces can be a terrible drain on the world's resources. Super-busy business people don't always put the environment at the top of their list of concerns, and many opportunities to save energy aren't taken because people don't think it's part of their job.

If you are the boss, finding time to introduce measures to cut your company's use of energy is well worth it, and can save you a lot of money.

Being seen to be green is also great business sense. With more eco-conscious clients and customers these days, having an environmental policy to put in your annual report (backed up by some real green credentials) can help win new business and impress your existing customers.

If you're not in charge, coming up with a green initiative and proposing energy-saving measures could help your career as well as the environment. It's always good to be the one who saves the company money on its bills and helps to improve its image.

At your own desk, make sure you set your computer up to save energy in the same way as your home computer (see tip 36). If you have an IT department, ask them to send round a note to your colleagues asking them to do the same thing. If you don't have an IT department or they are too busy, why not offer to write a note yourself?

To use a bit of dodgy business jargon, employing a bit of blue-sky thinking to be greener in the workplace is a win-win situation for everyone involved!

44 GET AN ENERGY AUDIT

The Carbon Trust has been set up by the UK government to help businesses cut their carbon footprints.

They will carry out a free, detailed energy survey for any business with energy bills over £50,000 per year, and their experts can suggest a wide range of low-cost actions to take.

For smaller businesses, the Carbon Trust website has a range of online tools to help assess energy use and put together an action plan to make savings. Their benchmarking tool also shows how your company compares with others of a similar size and type, so you can see how well you are doing.

They also provide notices and stickers to place around the workplace to remind employees to switch things off. The average office wastes £6,000 a year just from equipment left on outside working hours, so a few reminders here and there could mean significant savings.

Hiring one of the many environmental consultants out there can also pay for itself in reduced bills, if you want a detailed assessment of your company.

And don't forget to think about transport related to the business as well. Could your company invest in video-conference technology and save money on flights, as well as vast amounts of carbon?

Incentives to encourage employees to travel to work by bike or public transport are a great way to spread the green message. For business trips, why not suggest that travel expenses are paid by the mile for all modes of transport (as they tend to be for car travel), so that people claiming expenses for driving to meetings can claim the same amount even if they buy a rail ticket or go by bike.

45 PRINTERS & COPIERS

Laser printers and photocopiers both use high temperatures to seal the toner once it has been transferred to the paper. When left in print-ready mode, these machines can use up a lot of energy keeping warm.

Because printers and photocopiers are usually shared between several workers, they suffer from the classic 'tragedy of the commons' effect, where no one takes responsibility for turning them off, and they end up being left on twenty-four hours a day.

The simplest way to make sure these machines aren't left guzzling energy after everyone goes home is to get permission to put a prominent notice next to each one.

Something like 'If you are the last one here and this is your last bit of copying, please switch me off!' should help to reduce the number of times this happens in the future.

Other ways to reduce the energy used in printing (and save paper, too) include:
• Encouraging employees to add a footnote to their emails saying 'Please only print this email if absolutely necessary'.
• Replacing multiple desktop printers with one multi-purpose printer, fax and photocopier for every ten or so workers – the short walk this introduces when people need to pick up a printed document will reduce the temptation to print out absolutely everything that comes in, and will only leave one machine that needs to be switched off at the end of the day, too.

LOW-ENERGY LIGHTING 46

For health and safety reasons, workplaces must be well lit and contain a lot of light bulbs, so there's no excuse for not using energy-saving fluorescent tubes and light bulbs to illuminate your office, workshop or factory. These save a fortune in maintenance as well as electricity bills.

If your workstation is a bit dim and you find you need a desktop task light, small LED lamps can be a good investment, as they use a tiny amount of electricity and are ideal for this purpose.

47 NEITHER HOT NOR COLD

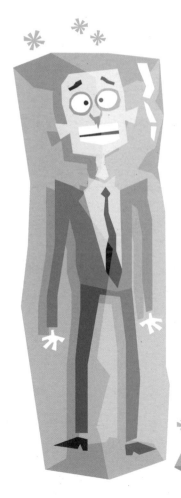

If your office has an air-conditioning/heating dial, make sure this isn't set too hot or too cold for the season. If everyone is shivering in a jumper in the middle of summer or wearing short sleeves in January, then things probably need adjusting.

A workplace without air conditioning can get very hot in summer. You can open windows strategically to maximise air flow without creating disrupting gusts of wind. Leaving a small gap at the top and bottom of a sash window can make a real difference to the temperature without sending the contents of your in-tray flying across the room every few minutes – and without everyone needing a fan next to their desk.

Other kinds of workplaces suffer from different problems. A garage or workshop may be open to the elements some of the time and get very cold in winter. Rather than heating up the whole space with a gas heater, infrared directional lamps can provide heat just where people are working and save some of the wasted energy.

LONG-TERM STRATEGIES 48

Some energy-saving measures, such as fitting double-glazing or changing light fittings, can only be done when a workplace is being refurbished.

Ensure that saving energy is a priority when these opportunities arise by encouraging your boss to produce a long-term strategy for energy saving and carbon dioxide reductions in your company.

This document is an excellent thing to be able to point to when people want to know what your business is doing to help the environment, and putting energy-saving goals on record will help to make sure these factors are taken into account when refurbishments and new procurement contracts are being considered.

SAVING ENERGY WITH OTHERS

Ranging from the simple to the very ambitious, joining together with others to organise and promote energy saving can be much more effective than struggling along on your own. The savings can really mount up once everyone is supporting each other, and it's fun, too.

49 ONE STREET AT A TIME

Getting work done to improve your home at the same time as others in your street can make a lot of sense.

Workers who need to travel to your street, set up equipment or scaffolding and order parts will make huge savings on labour if they can do more than one house at the same time.

If you are thinking of having your cavity walls filled, why not see if your neighbours have the same idea? If you can call round for quotes with several jobs on offer, you might get some great deals.

If the houses in your road all have south-facing roofs and you think solar water heating might be viable for you, then it will probably be cost effective for everyone else on your street as well. Rather than go it alone, a great thing to do is ask a reputable company to give a presentation about their solar water-heating products and then invite your neighbours to come along for an interesting social evening.

With several potential customers, you will probably find a good company is willing to make the effort to do this. And even if everyone doesn't take the plunge along with you, consulting and involving your neighbours in your green makeover will make it much easier if you have to ask for planning permission later on.

GET THE TOWN INVOLVED 50

The village of Ashton Hayes in Cheshire is famous for setting itself the goal of becoming the first carbon-neutral village in England.

Starting with a public meeting in 2006, residents have made ambitious plans to tackle energy waste in all aspects of village life, and to support each other's efforts to save energy and generate green energy locally.

So far, they have:

- Collected data from each house in the village to work out their starting carbon footprint (slightly above average).
- Made plans and got funding for a safe new footpath to the station to help get people out of their cars.
- Got planning permission for a wind turbine at the local school.

Several new sets of solar hot-water panels have gone up on local homes, with householders encouraging others to call in and see how they are performing.

The local pub is also aiming to be carbon neutral, and is starting with local ingredients for its meals, switching its electricity supply and planning to put in solar hot water.

Get inspiration and follow their progress at:
www.goingcarbonneutral.co.uk

FURTHER INFORMATION & ADVICE

WATER

Saving water in the home
The Waterwise website is a fantastic online resource for water-saving tips.
www.waterwise.org.uk

The Environment Agency is part of the UK government and has lots of information about conserving water resources.
www.environment-agency.gov.uk

The Water Guide is a hub for information about the UK water industry, with advice for consumers on where our water comes from, as well as tips on saving water.
www.water-guide.org.uk

Get simple water-saving devices for your loo from:
www.hippo-the-watersaver.co.uk
www.save-a-flush.co.uk

Hidden water
Calculate your water footprint using the handy tool on this website.
www.waterfootprint.org

Waterwise has produced a detailed report on the hidden water in the things we buy and the food we eat. Download it from the website.
www.waterwise.org.uk

Cutting down on waste helps to save hidden water in the things we consume. The Waste & Resources Action Programme campaigns to reduce packaging as well as increase composting and recycling.
www.wrap.org.uk

WasteOnline has factsheets on numerous topics surrounding waste and recycling, with tips on recycling everything from car tyres to computers.
www.wasteonline.org.uk

The Recycled Products Guide has a searchable database of products made from recycled materials.
www.recycledproducts.org.uk

Being more vegetarian

The Vegetarian Society provides lots of tips to help you eat less meat, as well as a host of good recipes.
www.vegsoc.org

The BBC Food's recipe archive has many excellent vegetarian and vegan recipes.
www.bbc.co.uk/food

The 8th Day Co-op Café in Manchester has collected together its favourite vegetarian dishes.
www.eighth-day.co.uk

The Savvy Vegetarian website has some great ideas for kid-friendly meals.
www.savvyvegetarian.com

Saving water in the garden

The Royal Horticultural Society has downloadable reports on many aspects of water conservation.
www.rhs.org.uk

Sunshine Garden has ideas for creating urban gardens that aren't thirsty.
www.london.gov.uk/sunshinegarden

Living roofs

The Living Roofs website has almost everything you need to know about building a green roof, including where to find expert advice.
www.livingroofs.org

Find out about the benefits of green cities and how to make the most of greenery on and around your house from the London Assembly Biodiversity Strategy report 'Building Green'.
www.london.gov.uk/mayor/strategies/biodiversity

ENERGY

Insulation, heating and cooling

The National Insulation Association has lots of information on how to improve the heat-retention ability of your home.
www.nationalinsulationassociation.org.uk

The Sustainable Building Association has factsheets covering a range of eco-building and maintenance issues.
www.aecb.net

George Marshall of the Climate Outreach and Information Network has been turning an ordinary terraced house in Oxford into a DIY eco-home for several years. Read the story and get loads of useful information from his website.
www.theyellowhouse.org.uk

The British Fenestration Rating Council can help you find the right kind of energy-saving windows for your house and put you in touch with a local supplier.
www.bfrc.org

The Energy Saving Trust has advice on energy saving and links to available sources of grants to help with insulation, as well as information on energy-saving heating.
www.energysavingtrust.org.uk

Warm Front provides grants to pensioners and people on benefits to insulate their homes.
www.warmfront.co.uk

The HEAT project website provides a clear and simple guide to grants and support for home energy saving and insulation.
www.heca.co.uk

The Log Pile website helps you to source locally produced, sustainable wood fuel for stoves and boilers.
www.logpile.co.uk

Green electricity

The Centre for Alternative Technology in Wales has been studying and testing green energy for years. They have a wealth of advice on energy saving and green electricity.
www.cat.org.uk

The Renewable Energy Centre provides information on the different green energy technologies.
www.therenewableenergycentre.co.uk

The Green Energy Works website from the Green Party shows examples of green electricity being generated today in homes, schools, workplaces and many other situations.
www.greenenergyworks.org.uk

The National Consumer Council produces reports to help people find their way around confusing topics, such as food miles and green electricity tariffs.
www.ncc.org.uk

The Low Carbon Buildings Programme administers national government grants for green energy projects in the UK.
www.lowcarbonbuildings.org.uk

The British Wind Energy Association promotes and monitors the development of wind energy.
www.bwea.com

Solar Century supplies solar panels to builders and DIY enthusiasts, and has interesting case studies of solar electricity projects.
www.solarcentury.com

Saving energy in the kitchen

The Energy Saving Trust has a guide to energy-saving appliances and an up-to-date list of recommended products and suppliers.
www.energysavingtrust.org.uk

The Women's Institute has launched a carbon challenge, and provides lots of simple tips for being greener everywhere in the home, not just the kitchen.
www.womens-institute.co.uk

Eco-friendly shopping

Ethical Consumer magazine rates a wide range of products on different ethical criteria. The website has sample guides available for free to non-subscribers.
www.ethiscore.org

Find your local farmers' markets on this website.
www.farmersmarkets.net

Get information on where to find the best vintage, recycled and ethical clothes, as well as details of greener food and beauty products on the Style Will Save Us website.
www.stylewillsaveus.com

Repair, reuse, recycle: find out how to remove bobbles on jumpers, repair zips and remove almost any stain, thanks to the huge mine of information collected on the Top Tips for Girls website.
www.toptipsforgirls.com

Greener products can be hard to find on the high street. Mail-order shopping can provide the answer, so use your search engine to shop around, or try these online shops to find a range of products more easily.

Centre for Alternative Technology shop
www.cat.org.uk/shopping

WWF's Earthly Goods shop has a wide range of eco-friendly products.
www.wwf.org.uk/shop

Nigel's Eco Store
www.nigelsecostore.com

Natural Collection
www.naturalcollection.com

Green Gardener
www.greengardener.co.uk

Saving energy at work
The Carbon Trust helps businesses reduce their impact on the climate.
www.carbontrust.co.uk

Saving energy with others
The town of Ashton Hayes has set up a website and blog to chronicle its efforts at becoming the first carbon-neutral town in England.
www.goingcarbonneutral.co.uk